MADE IN AFRICA

MADE IN AFRICA

*Learning from
carpentry hand-tool projects*

JANET LEEK ANDREW SCOTT
MATTHEW TAYLOR

VSO/IT PUBLICATIONS 1993

Practical Action Publishing Ltd
The Schumacher Centre
Bourton on Dunsmore, Rugby,
Warwickshire CV23 9QZ, UK
www.practicalactionpublishing.org

© Intermediate Technology Publications 1993.

First published 1993\Digitised 2013

ISBN 10: 1 85339 209 X
ISBN 13:9781853392092
ISBN Library Ebook:9781780442709
Book DOI: http://dx.doi.org/10.3362/9781780442709

All rights reserved. No part of this publication may be reprinted or reproduced or utilized in any form or by any electronic, mechanical, or other means, now known or hereafter invented, including photocopying and recording, or in any information storage or retrieval system, without the written permission of the publishers.

A catalogue record for this book is available from the British Library.

The authors, contributors and/or editors have asserted their rights under the Copyright Designs and Patents Act 1988 to be identified as authors of this work.

Since 1974, Practical Action Publishing (formerly Intermediate Technology Publications and ITDG Publishing) has published and disseminated books and information in support of international development work throughout the world. Practical Action Publishing is a trading name of Practical Action Publishing Ltd (Company Reg. No. 1159018), the wholly owned publishing company of Practical Action. Practical Action Publishing trades only in support of its parent charity objectives and any profits are covenanted back to Practical Action (Charity

Reg. No. 247257, Group VAT Registration No. 880 9924 76).

Contents

Foreword	vii
1 Introduction	1
An historical perspective	1
Re-introducing toolmaking	2
Barriers to success	3
A gender perspective	4
The text	4
2 Why make hand-tools?	5
Why carpentry?	5
Approaches to making tools	7
Summary	10
3 Hand-tools and practising carpenters	12
Practising carpenters	12
Carpenters' tools	14
Carpenters' own experience in making tools	15
The IT experience	17
The VSO experience	18
Conclusions	20
Lessons learned	22
4 Making tools in educational institutions	24
Benefits	24
Youth polytechnics, Kenya	25
Government training centres, Malawi	28
NGO training centres, Zimbabwe	29
Government training centres, Zimbabwe	32
Acceptability	33

	Toolmaking: an educational justification	34
	Conclusions	35
	Lessons learned	36
5	**Hand-tool production units**	**38**
	Rapogi youth polytechnic	38
	Bande Tool Makers	39
	WECO tool production unit	41
	Itawoga tools	43
	COMET, Zambia	45
	Conclusions	46
	Lessons learned	47
6	**Conclusions and future work**	**50**
	Skills transfer	50
	Appropriateness	50
	Acceptability	50
	Sustainability	51
	Combined programmes	51
	Which tools?	52
	Self-employment	52
	Further work by IT and VSO	53

References   55

Appendix 1: The tools   59

Appendix 2: Price comparisons of selected tools   71

Foreword

This publication should be read by many of the NGOs, bilateral and multilateral agencies which are currently attempting to promote income-generating projects for rural and urban micro-enterprises. It should also be of interest to Appropriate Technology units in African universities, and units mandated to promote informal sector enterprises in government ministries, parastatals and lending institutions.

What makes this publication refreshing is its awareness of the need to take time to effect change. It is also aware that technology adoption is not a technical question alone, but is inseparable from the values and attitudes of those working in small urban and rural industries. The West has been for so many decades the source of 'state of the art' technologies, often imposed in unsustainable ways in the developing world, that it is now quite difficult to accept that there is another West, actually designing and promoting something other than the latest technology.

VSO and IT are part of this Other West, concerned about affordability, sustainability and technological capability. In combination, they offer here an honest and realistic account of many years' work in the promotion of hand-tool technologies across eastern and southern Africa.

KENNETH KING
Director, Centre of African Studies
University of Edinburgh

1
Introduction

An historical perspective

Carpentry is a trade of relatively recent origin in most of Africa. It was introduced by European missionaries and colonialists to enable them to enjoy the style of buildings and furniture to which they were accustomed. Traditional woodworking in east and southern Africa employed, and continues to employ, carving techniques, producing products ranging from spoons and bowls to stools and head-rests, often ornately decorated. The tools used were locally-made adzes of various sizes and shapes, with blades at different angles, each designed for particular work. On the east African coast, home for several centuries to the rich Swahili culture, the builders of wooden trading and fishing boats and the makers of the ornate Swahili furniture and fittings made and used wooden tools.

A photograph of the young Jomo Kenyatta, taken in 1912, which appeared in the press on Kenyatta Day 1991, shows the future president of Kenya as a carpentry trainee, displaying a wooden jack plane that he had made. A carpentry teacher at a secondary school in Kenya's Western Province still uses a smoothing plane that his grandfather made at a local technical training institution in the 1930s. He had been taught by a European instructor. Another carpentry instructor in Kisumu uses a wooden plough plane which he bought from an old man, now deceased, who had worked for most of his life on the railways. These are just a few instances which illustrate that when the Europeans arrived to colonize Kenya they brought with them wooden carpentry tools, and the skills to make these tools were passed on to some Kenyan carpenters. European carpentry skills and tools were similarly introduced in other areas of Africa settled by colonizers.

The making of carpentry tools by carpenters is thus not a new phenomenon in east and southern Africa, but during the second half of the twentieth century the skills to make them have been lost by all but a very few. The European apprenticeship system, under which European wooden tools were developed and the skills for toolmaking were imparted, has not been transferred to Africa along with European-style carpentry. Instead, large foreign companies, or, in the case of Zimbabwe, local factories, have supplied the market with mass-produced, sophisticated metal tools.

Re-introducing toolmaking

In recent years the Intermediate Technology Development Group (IT), VSO and a number of other non-government organizations (NGOs) have begun to reintroduce the skills of toolmaking. This work has been taking place not only in Kenya, Malawi and Zimbabwe, where IT and VSO have mostly been active, but also in Botswana, Zambia, Tanzania, Uganda and Mozambique. But what has prompted this revival of toolmaking? If locally made tools have been displaced by factory-made tools, and toolmaking by carpenters never really took hold, what is to be gained now by making tools locally?

Kenya has the highest population growth rate in the world (at 3.7 per cent a year), and many other developing countries have a similar population growth problem. In such circumstances, as explored further in chapter 2, additional employment must be found, and enhancing the chances for self-employment seems a sensible option for many developing countries. Among the options for self-employment are small-scale carpentry businesses, a trade which appears to offer good opportunities. The skills and basic tools needed to work as a carpenter are relatively low-cost, and as incomes increase the demand for more, diverse and sophisticated furniture and fittings is likely to grow.

Many young carpenters, however, struggle to buy the tools they need to start a business. Practising carpenters often work with only the bare minimum of hand-tools. The imported metal tools that superseded locally-made tools in Kenya and elsewhere are expensive and beyond the reach of many carpenters. In many African countries, hand-tools, particularly specialist tools, are still quite scarce. The prices of imported tools are rising quite rapidly as a result of changing exchange rates, and the accessibility of such tools is likely to get worse in coming years.

It therefore seems a logical and sensible solution to teach carpenters to make their own tools, as well as to encourage local commercial production. This would have the additional benefit of allowing carpenters to repair their own tools, of building confidence in things made locally and an attitude of self-reliance. If successful on a large scale, it would reduce the need for valuable foreign exchange for the import of tools.

The advantages of making tools locally with the strategies for doing so are presented in the following chapters. The local manufacture of hand-tools, for personal use and for sale, is examined. Depending on local circumstances, these tools might be made wholly of wood, such as mallets and gauges, some might be made of wood with metal parts or blades, such as planes, chisels and hammers, while others might be made entirely of metal, such as planes. The choice of materials will depend very much on the cost and availability of the necessary materials, hardwoods and metal (scrap or new). Although a few small-scale carpenters have powered tools, such as lathes, and some are known to have made their own, the text concerns itself primarily with hand-tools, the basic equipment of a carpenter. (Appendix 1 describes these tools in more detail.)

The experiences of IT and VSO in training carpenters in toolmaking are described in chapter 3, and in chapter 4 the introduction of toolmaking to educational and training institutions is presented. Chapter 5 describes the establishment of tool production units which use electrical machinery for the manufacture and supply of hand-tools in larger numbers than can be made by hand.

Barriers to success

Despite the fact that the solution of training carpenters to make tools sounds simple, almost obvious, those involved in toolmaking projects in recent years have encountered many barriers to success, not least being the acceptability of these tools. Many carpenters regard wooden tools as a technological step backwards. Even when they work perfectly well, locally-made tools, particularly the wooden ones, look less sophisticated, less modern, than their factory-made equivalents. This has undoubtedly led to a reluctance on the part of today's carpenters to use them. The problems encountered in introducing toolmaking are therefore also explored in the following chapters. The lessons which have been learnt in each of the three approaches are summarized at the end of each chapter as an aid to others contemplating similar projects.

A gender perspective

Carpentry in Africa is predominantly a man's occupation. While a woman carpenter was encountered at the Kisumu *jua kali* during a survey by VSO, and Glen Forest Training Centre in Zimbabwe has trained several women carpenters, these are the exceptions rather than the rule. In more conservative Malawi, women carpenters are unheard of. For this reason all of the examples and individuals referred to in the text are men.

The text

The following chapters focus on the experiences of IT and VSO in carpentry hand-tool production in Kenya, Malawi and Zimbabwe. These experiences have encompassed teaching toolmaking in technical training institutions, teaching practising carpenters and the establishment of hand-tool production units. While IT and VSO have been working independently in Africa, it might be noted that the same person, Aaron Moore, began the work on tools in each organization. Another link between the two organizations has been the training in carpentry toolmaking for VSO volunteers held regularly by IT since 1989. A total of 37 volunteers have received this training prior to their postings, many of them being posted to Kenya and a few to Zimbabwe.

2
Why make hand-tools?

Why carpentry?

In each of the countries of eastern and southern Africa, the majority of the population lives in rural areas. For most of these rural households, non-agricultural activities contribute a significant proportion to family incomes, and for many they are the principal source of employment and income. In towns and cities, small-scale enterprises and the informal sector are at least as important a source of employment and income as the formal sector.

In Kenya, over 700 000 people are employed in small-scale enterprises and the informal sector, collectively called the *jua kali* (which in Kiswahili means 'fierce sun', thus depicting artisans working in the open). In Zimbabwe, where the term *dura wall* is often used to describe the informal sector (after the pre-fabricated concrete wall bordering their working area in Harare), a recent survey estimated that over one million people are wholly or partly engaged in small-scale enterprises. (McPherson, 1991) In Malawi, small- and medium-scale enterprises and the informal sector employ over 450 000 workers (READI, 1987). Clearly, small-scale non-farm enterprises are of great importance as a source of employment for large numbers of people.

There is now widespread recognition among government planners, development agencies and the community at large, that promotion of self-employment and wage employment in small-scale enterprises can contribute significantly to the local and national economy. Increases in employment to date have not kept pace with population growth, and today hundreds of millions of people do not have secure or adequate employment. By the end of the century over one billion more people will have joined the world's labour force. Traditional agriculture and formal sector employment are unable to absorb this number of people. In Zimbabwe, to take one example, over 100 000 school leavers seek employment every year, but there are fewer than 30 000 formal sector jobs available for them. Employment creation in the so-called informal sector is therefore seen as a means to ease this critical employment problem.

In addition to the potential number of workplaces, expanding employment in the small-scale enterprise sector offers other advantages. The creation of a new job in the informal sector costs less than in the formal sector. Enterprises in the informal sector tend to be labour-intensive, consistent with the abundance of labour in developing countries, and typically cater for the very poor by creating employment opportunities and security for the extended family and community members. The informal sector conserves scarce foreign exchange as it does not rely on large-scale imported machinery.

Non-agricultural economic activities can take many forms, ranging from handicrafts and small-scale industry to personal services and retailing. Carpentry, for which only minimal capital investment and basic skills are required before operation can begin, is one of the commonest types of small-scale manufacturing enterprise throughout eastern and southern Africa. In Kenya, 4–5 per cent of all informal sector employment is in carpentry, while in Zimbabwe wood-processing enterprises are the third most common type of small enterprise. (Livingstone, 1991; McPherson, 1991)

Small-scale carpentry enterprises, whether in the formal or informal sector, are therefore currently a means of livelihood for a significant number of people in eastern and southern Africa. An increasing number of school-leavers lack the knowledge and skills to meet labour market demands and have no further training opportunities. Promotion of employment in the small-scale enterprise sector, including carpentry, through skills training and supply of equipment, offers the potential for creating this employment. Increasing the availability and reducing the costs of the basic equipment to enter employment or self-employment will increase the income-earning opportunities for those with the basic skills. For carpenters this means the provision of hand-tools, locally-made and at lower cost than the factory-made tools presently available.

Approaches to making tools

There are three avenues open for the introduction of locally made hand-tools:

o by training practising carpenters to make their own tools;
o by training carpentry trainees and new entrants to the trade to make tools; and
o by establishment of tool production units.

Each of these approaches has been tried in some form in Zimbabwe, Kenya and Malawi. The first strategy leads into areas of small enterprise support, and in particular skills training for existing businesses. The second entails the training of instructors and the changing of training curricula within educational and vocational training institutions. The third also involves business development. The different objectives which can be achieved by each strategy are outlined in the remainder of this chapter and are summarized in table 1.

Local hand-tools in small-scale enterprises

In many developing countries even the simplest of carpentry tools are expensive, and specialized tools (e.g. plough planes) are often unobtainable. In Kenya, where carpentry tools are quite widely available, the prices put many, particularly the specialized tools, out of the reach of many small-scale carpentry businesses. In Malawi, for example, investment in a basic toolkit costs the equivalent of twice the annual turnover of a rural carpenter. (Scott, 1987) Information about actual tool prices is presented in appendix 2. Aaron Moore, formerly a VSO volunteer in Kenya and IT Project Manager, realized that 'a lot of craftsmen being trained would not be able to afford the tools, and having the skills but not the tools is criminal.' (*The Sunday Times*, 9 September 1991) Enabling practising and established carpenters to buy and use low-cost, local tools as well as teaching them to make tools for themselves would go some way to overcoming this problem. With adequate tools, small-scale enterprises would be able to maximize their potential without a large capital outlay.

The promotion of locally-made tools amongst practising carpenters can help achieve the following objectives:

o To provide carpenters with access to quality local tools that are cheaper than the imported ones and to make local tools available where imported ones are scarce.

- To create confidence in locally-made goods. This applies when practising carpenters are seen by their fellows to be using locally-made tools which they have purchased, and when carpenters make their own.
- To enable carpenters to make and repair their own hand-tools because it is much cheaper to make and repair tools than to buy them, and it gives them the confidence to solve problems for themselves.

Local hand-tools in educational institutions

Within the education systems of Kenya and Zimbabwe, and indeed in other countries, an element of practical carpentry is required in both primary and secondary school syllabuses. To meet these needs each school should have kits of basic carpentry tools, but in reality most primary schools and many secondary schools are unable to afford such tools. Vocational training centres for school-leavers face the same problem. In Zimbabwe, the government-run youth training centres were established to provide training for primary school-leavers, with a view to formal sector employment. Though government institutions they do not all have sufficient tools for their trainees.

Similarly, many NGO training centres are severely under-resourced and do not have the equipment to provide the training they would like. Kenya has technical training institutions known as youth polytechnics which were set up in 1966 with the aim of preparing primary school-leavers for self-employment. Youth polytechnics are largely locally funded, and offer a variety of courses in different trades including carpentry. Although many are equipped with basic tools, they are often insufficient for the numbers of students accepted. Graduates of youth polytechnics often struggle for many years to obtain a basic tool-kit.

The promotion of toolmaking in the formal educational and training systems has two main aspects; firstly the training of instructors in toolmaking, and secondly changing the training curricula of individual institutions and nationally. The objectives which can be met through involvement in this area are as follows:

- To provide educational institutions with affordable, quality and durable tools, in order to facilitate carpentry training.
- To create a confidence in things made locally. Henry

Rugendo of ActionAid, Kenya explained that many Kenyans believe that a shirt 'Made in England' is far better quality than a similar shirt 'Made in Kenya'. *'It is therefore necessary to develop a culture that good things don't have to come from outside.'*

o To teach carpentry trainees to make hand-tools thereby providing a valuable learning experience. All over Kenya carpentry trainees learn to make sample joints out of small pieces of wood that are then simply discarded. Making hand-tools uses the same skills. When a sample joint is made to be thrown away it does not really matter if it is well done, but when a tool is made it has to be good to work, and the students have made something they can use.

o To provide carpentry students, who have made their own tools during training, with a medium-sized tool-kit to take away on graduation in order to enhance their chances of employment. A VSO survey of youth polytechnic and *jua kali* artisans, conducted in 1990, indicated that over 70 per cent of trainees wish to enter the *jua kali* sector but 65 per cent will join this sector without the necessary tools with which to practise their newly-acquired skills and earn a living. Formal sector employers also frequently expect carpenter employees to provide their own tools.

o To teach carpentry trainees toolmaking in order that in the future, as carpenters, they will have the necessary skills to make the tools they might require and the confidence to look for ways of solving 'problems' without having to rely on expensive imported 'solutions'.

o To teach other carpentry instructors in educational institutions how to teach toolmaking to their own students.

Carpentry hand-tool production units

Besides their cost and availability, as already mentioned, another problem with imported tools is their quality. In the case of cheap Chinese imports which are generally made for non-professional (DIY) use on soft woods, the tools have a very limited life when subjected to intensive use which includes work on tropical hardwoods. The bodies of many such tools are constructed of brittle white metal which cannot be repaired after breakage.

Tool production units can address these problems. They can

range in size from a single artisan making some basic tools for sale in the locality, to a formal sector manufacturing business making tools for the open market and competing directly with the well-known imported brands.

Establishment of tool production units can achieve the following objectives:

o To produce a range of carpentry hand-tools that are cheaper than their imported equivalents.

o To produce a range of tools that are, in terms of quality and durability, at least as good as their imported equivalents.

o To make available those tools that are otherwise scarce or hard to obtain in that locality.

o To reduce the need for foreign exchange for carpentry hand-tool imports.

o To create employment.

Summary

It should be realized that these three approaches are interrelated. For example, practising carpenters can be trained at training centres, and they can buy tools from production units. The youth polytechnic graduate can set up a small-scale business, make his own mortise gauge and sell one to a friend. The principal market for tool production units could be the educational institutions.

Table 1 summarizes the objectives which toolmaking helps to achieve and shows which strategy is likely to meet each objective. Although the objectives of toolmaking under each approach are distinct, they overlap and can help attain the same overall aim of increasing employment and income-earning opportunities for carpenters. It should also be noted that in deciding, for example, to introduce toolmaking to educational institutions, not all of the listed objectives need be tackled. The extent to which each objective is achieved will depend on the circumstances of the programme concerned.

Table 1: Toolmaking strategies and their relevance to different objectives

Objective	Strategy		
	Educational institutions	Practising carpenters	Production units
To provide low-cost, readily available hand-tools	M / L	M / L	H
To produce good quality, durable tools locally	H	M	H
To improve training in carpentry techniques	H	M	L
To give confidence in locally-made tools	H	M / L	L
To teach people to make and repair their own tools	H	M	L
To give confidence to solve own problems	H / M	H / M	L
To equip training institutions with hand-tools	H	L	M
To reduce need for foreign exchange	H / M	M / H	H
To increase chances of self-employment	H	L	L

H = High relevance   M = Medium relevance   L = Low relevance

3
Hand-tools and practising carpenters

Practising carpenters

There is great diversity amongst small-scale carpentry enterprises. Surveys by IT, VSO and others have generated information about both rural and urban carpenters in Malawi, Zimbabwe and Kenya, which have shown that they differ in scale, in skills, in markets and in level of investment. One of the clearest differences is between urban and rural carpenters. It is therefore difficult to generalize when describing the characteristics of carpenters.

Ten artisans, all members of *jua kali* associations, were interviewed by VSO at three locations in western Kenya, Kabras, Kakamega and Kisumu. *Jua kali* associations have been instigated by the Ministry of Technical Training and Applied Technology with the primary aim of registering and facilitating loan distribution to artisans in the informal sector. The associations are also distributing donated land, facilitating training and setting up showrooms to display *jua kali* work.

At Kabras, the *jua kali*, based in a small rural village, is one of the smallest with only 64 members, 12 of whom are carpenters. The members are spread out through neighbouring villages. At Kakamega, a small town about 30 kilometres from Kabras and 55 kilometres from Kisumu, the workshops are all centrally located, in an area near the bus park, on land belonging to the association there. The furniture being made by carpenters in Kakamega (i.e. standard tables, chairs, cupboards and beds) is only slightly more sophisticated than that being produced in Kabras.

The Kisumu *jua kali*, with a membership of several thousand, presents a completely different picture. Kisumu is Kenya's third biggest town, where the carpenters are centred in a large market place and occupy row upon row of shacks, densely packed together. Nearby, a large group of metalworkers occupy another section of the market. The type of furniture being produced is far more diverse and sophisticated than at Kakamega and Kabras. The quality of finishing and joints is also noticeably better. The Kisumu consumer obviously has more money and more sophisticated tastes.

Mombasa is Kenya's second city and one of east Africa's oldest settlements, with an advanced Swahili culture dating back many centuries. Consequently, relatively sophisticated carpentry techniques introduced by the Arab settlers also have a long history here. As traders and fishermen the Swahili people built and used Arab-style wooden boats called *dhows*, and their houses displayed ornately carved and sophisticated fittings and furniture. In Mombasa the majority of carpentry workshops surveyed by VSO were owned by older men, housed in larger premises and employed more people than in the locations visited in western Kenya. The furniture being produced was, as in Kisumu, diverse, sophisticated and of high quality, possibly more so. Traditional Swahili carving was incorporated into many of the designs.

By contrast, in Zimbabwe and Malawi rural carpentry is generally a seasonal activity which allows continued involvement in agriculture while providing a seasonal source of cash income. Production is primarily of low-cost items whose acceptability to customers is based on low price rather than the quality of work or materials. The products comprise basic, cheap furniture and, in many cases, building joinery (i.e. roofing, doors, window frames). The level of output of each carpentry unit is quite small. Revenues earned from carpentry are used to meet essential cash requirements but are not the sole source of support in most cases. (Cromwell and Moore, 1988) For most rural carpenters, carpentry is a secondary source of income for their families and to describe them as 'entrepreneurs' or 'businessmen' would be misleading.

There is great variation between carpenters' skills. IT found in Malawi that only a quarter of all carpenters had had formal training. In Kenya the great majority of carpenters had been trained 'on the job' by other carpenters and had not attended any technical training institution. In Zimbabwe a third of all rural carpenters had been trained through informal apprenticeships. (Cromwell and Moore, 1988)

Carpenters' tools

The tools that practising carpenters own vary from enterprise to enterprise. Almost all have a jack plane, saw and clamp, as well as a hammer and chisel but beyond this there is great variation.

The artisans surveyed by VSO in Kisumu, particularly those who are self-employed, owned a much wider range of tools than those in Kakamega. All the artisans interviewed had bought their tools gradually over a period of time by saving their wages and profits. One carpenter at Kisumu produced a wooden plough plane that he had bought from an old man who had worked on the railways. The artisans shared tools, especially specialist tools that are expensive to buy, and many of the other carpenters borrowed this plough plane and felt that it worked better than the imported version.

In Mombasa there was very little evidence of the wooden tools of the past which have been superseded by imported tools. The *jua kali* workshops are relatively well equipped with hand-tools, and when asked what additional tools they would like to expand their workshop, the overriding need expressed by the craftsmen was for machine tools such as lathes. A local craftsman producing extremely high-quality traditional furniture said that the only locally-made tools he was willing to buy were very specialized carving chisels now made by only one craftsman on Lamu, an island further up the coast.

The survey of *jua kali* by VSO, and the initial survey in Zimbabwe for IT, both entailed demonstration of locally-made carpenters' tools. Reactions in Zimbabwe were reasonably positive, and most of those interviewed indicated that wooden tools would be bought because they were cheaper. In particular the more sophisticated planes (plough and rebate planes), which are very expensive, were of greatest interest in both Kenya and Zimbabwe.

Three carpenters, working for a businessman at Kabras and turning out basic furniture with simple designs, were given a WECO (Western College of Arts and Technology) metal jack plane and plough plane to try out. When visited the following day, they had tried the plough plane and said it worked well. The jack plane they had hardly tried at all, citing the problem as blade-setting. When encouraged to try again, they managed to set the blade with virtually no assistance and, putting the jack plane to use, they found it worked well. They were still sceptical, however, one observing that it was too long, though in fact it was just a different standard size to the one they used. Another complained that it had square, not round, corners.

None of the three owned their own tools; the jack plane cost about a third of the price of the quality imported equivalent, and the plough plane about a fifth of the price. All three carpenters expressed an interest in being self-employed but said their lack of tools prevented them. Their reluctance to accept locally-made tools, despite a clear cost difference, demonstrates that there are other factors to be considered when a tool is purchased. This problem of acceptability is discussed further in chapter 6.

The response to the tools at Kakamega was quite different. The tools demonstrated were again made at WECO, and included metal and wooden jack planes and a rebate plane. There was enormous interest in the planes, with the metal jack plane receiving the most interest. Other artisans gathered around to watch the trial, and nobody present had seen locally made hand-tools before. The artisans testing the tools expressed an interest in buying the jack plane, as did some of the observers. All three participants employed other carpenters and needed additional tools to enable all of them to work at the same time.

Carpenters were also asked about the tools they had and the tools they needed to expand their business. In Zimbabwe, all but three out of thirty carpenters felt that they were under-equipped. The most sought-after tool was the grooving (plough) plane. Others sought were jack planes, rebate planes, saws, vices, chisels, spokeshaves and try squares. At Kakamega, artisans mostly wanted more of the same tools so that they could have more employees working at one time. The Kisumu artisans mostly wanted tools that they did not already possess in order to expand the diversity of their tool-kits.

In Malawi, a survey covering a total of 43 carpentry enterprises found the lack of tools and the care of tools to be important problems for carpenters. In all, almost a quarter of the carpenters had a problem with lack of tools. An earlier survey in Lilongwe region found that 14 per cent of carpenters claimed lack of tools to be a problem. (Scott, 1987)

Carpenters' own experience in making tools

Many carpenters have made, or have attempted to make, their own tools to overcome the problem of lack of tools. During the

Zimbabwe survey by IT in 1988, when 30 carpenters were visited, more than a third had made tools themselves. These included marking gauges, mallets, chisels, try squares, brace and bits, wooden sash clamps, saw sets and lathes. In addition, some wooden repairs to tools had been made (saw handles, gauges) as well as metalworking/welding repairs. In Malawi, a survey of the Mulanje South area in 1988 also found several self-made tools. These included a number of wooden planes, a simple yet effective vice and a lathe.

There is also evidence in Kenya of indigenous toolmaking in response to the need to find a solution to the lack of tools. At Kisumu, a number of carpenters have tried making simple marking tools by copying the designs of the imported ones. The quality was poor but this was partly due to the use of inappropriate materials. Most of the artisans bought their chisels from a metalworker in the neighbouring section of the *jua kali*, and made their own handles. One carpenter produced a metal plough plane, also made by a neighbouring metalworker who turned out to be a young man, Maurice Ochieng, aged 19. He had learnt to make the plane at a local technical training institution and had so far sold four at about £5 each, less than 20 per cent of the cost of its imported equivalent. This plough plane worked very well.

Indigenous hand-tools were not very apparent at Mombasa, but locally-made lathes were very common. The chairman and secretary of the *jua kali*, along with a local businessman, were currently testing the market with a range of very professionally-made machines that they hoped to put into production. The lathe, for example, was going to retail at about 55 per cent of the imported equivalent.

Training practising carpenters to make their own tools can thus be seen as a logical step which builds upon carpenters' own experience of toolmaking, and strengthens their abilities to overcome the tool problem themselves. The indigenous manufacture of hand-tools, and in some cases powered tools, within the small-enterprise and informal sector, as opposed to factories in the urban formal sector, has not been widely recognized, partly because it is generally meeting only very local needs. IT and VSO have now had experience in Malawi, Zimbabwe and Kenya, in developing the skills to improve upon the existing small-scale production of hand-tools.

The IT experience

The first carpenters to receive instruction in toolmaking from IT were trained in Malawi in early 1987. By the end of the first year, over a dozen courses had been held and over 100 carpenters trained to make tools. Several courses for carpenters have since been held in Zimbabwe, and VSO has held two courses in Kenya.

In Malawi, it was found during follow-up visits that three out of four carpenters were using at least one of the tools made on the courses. Later evaluation found that there is no overall pattern to the kinds of tools which are being used: the individual needs of the carpenters and their circumstances dictated which were used. The carpenters in Malawi reported a gain in efficiency using wooden tools, with the time taken to make some products (e.g. panelled doors) shortened. This potentially enables more orders to be met during their production season. The quality of products was improved, and some carpenters reported that the wooden tools were easier to use. (Scott, 1987)

In Zimbabwe, IT's support to the training of practising carpenters has been more indirect, through the technical assistance being provided to NGO carpentry instruction, in particular at Silveira House and Glen Forest Training Centre (GFTC). At the latter, the first course for practising carpenters, in 1988, lasted three weeks, during which the participants made a wide selection of wooden tools, whether they were required by the carpenters or not. This course also included three days of blacksmithing, when plane blades and chisels were made. During an evaluation of GFTC carpentry training in 1990, two of those interviewed, James Mugaviri and Solomon Mhepo, had attended this course. In both cases, at the time of the evaluation the tools made had been lent to other local carpenters, but they were being used. Both of these carpenters were already well equipped before the toolmaking course, and had no real need to attend the course.

The second course at Glen Forest, in August 1990, was only one week in duration and the seven participants made only three tools (a sash clamp, a rebate plane and a plough plane). These tools were selected after taking account of the priorities of the participants. All seven were sponsored by the Zimbabwe Council of Churches, who intended to help five of them set up in business as the Chinaa Youth Project at Chipinge. The other two were practising carpenters from co-operatives near Gutu. Follow-up visits in 1991 found that the youth group had broken up when most of the members had found jobs in towns. One of the others had not made any tools since the course, though the Zviito co-operative to which he belonged had purchased tools. The effectiveness of the toolmaking courses in Zimbabwe, therefore, in contrast to Malawi, appears so far to have been limited.

The VSO experience

VSO in Kenya is currently involved in the implementation of the Youth Training Support Programme (YTSP), a programme designed by VSO with the Ministry of Technical Training and Applied Technology and funded by the European Community. The target group of the YTSP comprises graduates of youth polytechnics who are trying to establish themselves in self-employment. Three production workshops are currently reaching completion where youth polytechnic graduates will be able to attend one- to two-week courses on specific specialist areas within their trade. For example, masons can attend a course in ferro-cement techniques thus giving them sufficient skill in a specialist area to enable them to seek employment. Each course participant will also be invited to a course on relevant business skills and will receive follow-up visits from a VSO volunteer to help them put their new skills into practice.

Before the YTSP centres had been completed, a few preliminary pilot courses were conducted at adjacent youth polytechnics, including two courses in carpentry hand-tool production catering for a total of eight practising carpenters. Four carpenters, all youth polytechnic graduates, aged between 19 and 24, attended the course at Kianjai Youth Polytechnic, where they learned to make six tools (set square, G-clamp, bevel square, mortise gauge, mallet and rebate plane). Although the course was free, participants had to pay the cost price (i.e. cost of materials) in order to take away the tools they had made. The second course was run in Machakos District by another VSO carpentry instructor. Here the participants were allowed to take the tools they made free of charge.

Initial follow-up indicates some use of the tools but detailed follow-up work is yet to be carried out. One of the four carpenters on the Kianjai course, Evans Nderi, had already had some experience of toolmaking and he is currently setting up a tool production unit with support from Approtec and ActionAid (see chapter 5).

Another participant, Peter M'eringo, aged 19, left Kianjai Youth Polytechnic having failed to complete his training because of financial problems at home. Since leaving the youth polytechnic he has carried out a few contracts for other carpenters and has managed to accumulate some basic tools: a jack plane, one sash clamp, four chisels, a handsaw and a try square. With these and some knocked-together planks which constitute a work bench under a tree at his home, Peter has managed to set up his own business, making coffee tables, chairs and beds for people in the immediate locality.

After the YTSP toolmaking course, Peter immediately bought the rebate plane he had made (the most expensive of the tools), but did not buy any of the others. With his rebate plane he now gets orders for doors, and window frames which he previously would have found difficult to make. Peter commented on his rebate plane,

> *'I think it is good. Wooden tools have the advantage of being cheap to make and easy to repair if they break. It is easy to set once you know how and it is better because it uses a chisel and you don't have to buy blades for it.'*

The tool he would like most to improve his business is a plough plane, and he would like to learn to make his own as those available in the shops are very expensive. However, Peter was not particularly interested in buying the other tools he had made and continues, for example, to improvise using nails to hold his wood in place even though the G-clamp would have cost him less than the rebate plane.

A third participant, Joshua Mairiti, aged 24, has a small carpentry workshop where he makes all kinds of furniture and employs two other carpenters on a piece-time basis. Joshua was fortunate that, on leaving the youth polytechnic, he managed to buy some second-hand tools cheaply from the family of an old man who had died. He started working for an established carpenter and gradually saved up for the tools he would need to start his own business. This has now been running for about a year, and Joshua is now reasonably well-off for tools. Following the toolmaking course he bought the marking gauge, bevel square, G-clamp and rebate plane that he had

Joshua Mairiti using the rebate plane he made on a YTSP course.

made. Although he already had an imported rebate plane, he and his employees use the one he made when the other is being used. The other three were new tools to his tool-kit. Of the tools he made, Joshua said,

'They are good because they are cheap and the quality of work they produce is good but they take longer to use.'

Joshua has obtained an order to make 15 G-clamps, bevel squares, try squares and marking gauges for Kiangai Tool Making Centre and, although this is an NGO-backed order, it has prompted some interest from other local carpenters. Joshua has sold a marking gauge and a larger version of the G-clamp to fellow artisans. It remains to be seen whether Joshua will continue to make tools in the future.

Conclusions

It has been shown that rural carpenters can learn to make hand-tools to a satisfactory quality, that the tools will be used by the carpenters, and that in some cases additional tools will be made and the knowledge of making tools will be passed on to others.

The limited experience gained from VSO's training of artisans through the YTSP programme suggests that there may be potential for training carpenters in hand-tool production in Kenya. In Zimbabwe and Malawi the experience of IT suggests a cautious approach. The visits to the *jua kali* locations described above suggest not only a need for access to these tools but also an interest in learning to make them.

In Malawi, the benefits that the carpenters were found to gain from using these tools stem largely from the time-saving the additional hand-tools provide. The carpenter might be able to increase his total production by using the time saved to produce for more orders, as one or two carpenters in Malawi reported. Alternatively, the time freed might be devoted to other economic activities. In Kenya, however, the length of time required when using the tools was found, by at least one carpenter, to be longer.

The hand-tools can provide further benefits to carpenters, and their customers, through the improved quality and wider range of products which they make possible. With follow-up training, concentrating on use of the tools and on carpentry techniques, this type of benefit can be enhanced. Another qualitative benefit to carpenters, which is impossible to quantify, is the ease of use of the hand-tools in comparison with the alternatives. The hand-tools require less effort to use.

Lack of tools is only one of the problems experienced by Peter M'eringo and Joshua Mairiti. Amongst Peter's other problems, for example, is his need for a more suitable workplace with easier access to transport, materials and a wider market. Joshua, although a skilled carpenter, is very much in need of some basic business skills such as accounting and pricing.

The survey of rural carpenters in Zimbabwe by IT found that over 60 per cent had problems with cash flow and working capital. Twenty per cent indicated a problem with raw materials shortages, and 11 per cent a problem with a lack of tools. While clearly the problems of finance have implications for tools in that their high cost has led to non-replacement, lack of tools was again not the only problem, nor necessarily the most pressing problem, experienced by practising carpenters. Raw materials, both in terms of availability and quality, and working capital to purchase raw materials are severe problems. The limited purchasing power of the majority of the people is of course the major constraint on increased carpentry production.

Lessons learned

Diversity Within Kenya and Zimbabwe it is clear that carpentry in the informal sector is extremely diverse. It is less so in Malawi. It cannot therefore be assumed that carpenters in different locations will have the same needs, including needs for skills and tools. Careful project planning is therefore required, based upon local needs.

Target area It seems that the most likely place for the success of a toolmaking project is a town, preferably with a high density of artisans working close together. Artisans in urban areas were found to have a higher level of motivation, possibly because they have no *shamba* or farm to fall back on. There was greater openness to innovation and the close proximity of many artisans meant that ideas were shared.

Programme centres A toolmaking programme should be carried out as close to the target area as possible. Ideally the training environment should attempt to resemble the working environment of the artisans. Programme facilitators should work closely with the artisans, spending time getting to know their working practices and needs. Projects should be based on interaction and not handed down to artisans on the basis of perceived needs.

Course length If practising carpenters are to attend courses, programmers should bear in mind the difficulties, financial and otherwise, of an artisan leaving his workplace for an extended period of time. If the course centre is close by, it should be easier for the artisan to attend a course and keep an eye on his business.

Target age It is likely that the artisans who should be targeted are those who have spent a couple of years working for another carpenter regardless of whether they have had formal carpentry training or not. This sort of experience gives the artisan sufficient skills, knowledge of the business world and perhaps capital to maximize their use of hand-tool training. In Kenya, this would correspond to a minimum age of about 23.

Tools The tools that artisans are taught to make should be chosen carefully according to the needs and demands of each individual. New tools could be introduced alongside training in new skill areas that will help the artisans diversify their products. Tool designs should be as simple as possible. The simpler the designs, the more likely they are to be copied and the more likely a carpenter is to put aside the time to make them.

Metal parts Where artisans of different trades work in close proximity, as described in the Kisumu example, carpenters should be taught to make the wooden parts of the tools whilst metalworkers should be taught to make the plane blades and other metal tools and parts.

Other needs It was pointed out in the introductory part of this chapter that the needs of the informal sector artisans are many and diverse. Whilst tackling the problem of tools, other needs must also be tackled in the same or a parallel project. Only when a diversity of needs are met can self-employment possibilities be best enhanced.

4
Making tools in educational institutions

Benefits

While many practising carpenters, in some areas the majority, acquired their skills outside formal training systems and institutions, it is also true that for an increasing number of carpenters entering the informal sector, formal training has been the source of their skills. Many youth polytechnic graduates, for example, now operate successful businesses in Kenya. Many of those employed in the formal sector received their basic skills in the educational system. Increasing numbers of practising carpenters have access to skills-upgrading courses at training centres.

The introduction of toolmaking to the educational and vocational training system can provide three main benefits. Toolmaking in a training institution enables it to equip itself at much lower cost than if it had purchased tools. It enables graduates, when they complete their training, to take with them a set of tools with which they can start earning a living, either in self-employment or working for someone else, and which they can replace in the future. It introduces prospective carpenters to locally-made wooden or metal tools at the same time as they themselves are introduced to carpentry, thus familiarizing them with such tools from the start. In the words of Peter Gilbert, VSO Field Director in Kenya,

> *'It equips the institution and demystifies the technology. It widens the so-called syllabus and where there is no syllabus it widens the training spectrum. Finally, it allows the trainees to carry something from the institution which they themselves have made and can use in their own future.'*

The introduction of toolmaking to educational establishments has been pursued in different ways, though each has required trained instructors and changed syllabuses. VSO volunteers posted to educational institutions have attempted to bring about change working from within individual institutions. IT has worked by training instructors and providing technical assistance from outside.

Youth polytechnics, Kenya

In January 1983 Aaron Moore was posted as a VSO carpentry instructor to Kaaga School for the Deaf in Meru District, Kenya. He arrived to find a workshop without tools and realized that if he equipped it with imported tools this would not help the students after they had left. So he determined on the simple idea of making tools at the school. Aaron had the skills as he came from a classical carpentry background, and he had a good knowledge of traditional woodwork tools. Therefore he started producing his own tools and from that began the concept of hand-tool production in VSO. Aaron produced some simple designs for carpentry tools that could be made from wood and scrap metal, which have since been published (Moore, 1986 and 1987), and with these introduced the concept to IT.

Aaron's 'simple' answer has been introduced into some 15 VSO projects in educational institutions around Kenya where similar problems of a lack of tools have been faced by volunteers. The majority of projects where carpentry hand-toolmaking has been introduced by VSO in Kenya are the poorer youth polytechnics, which are mostly reliant on the trainees' fees and community assistance for their survival. The financial constraints and the realities of day-to-day operation in such situations often makes the introduction of 'simple' ideas a little more complex. Another VSO carpentry instructor noted that the lack of motivation due to the low salaries of instructors was also a source of frustration and sometimes led to the private use of materials and tools by instructors as compensation for their poor pay.

Since toolmaking is not part of the youth polytechnic syllabus, volunteers have had to find spare time for it. Where spare time is available, it is often regarded as better used if easily-sold items, such as simple furniture, are produced as a means of income-generation for the polytechnic. Therefore, the ability of an individual volunteer to introduce a 'simple' and, on the face of it, sustainable idea to meet an obvious need is often problematic. To get toolmaking going in such institutions involves more than just technical know-how.

Whilst all VSO carpentry instructors who had taught toolmaking felt that their students had gained from the experience, none could claim much success. Perhaps the best example of limited success comes from Mulango Youth Polytechnic where Peter Wilkinson helped over 100 trainees to leave the polytechnic with a basic tool-kit in the two and a half years he was there. Of these, a couple had made some tools for other carpenters and a few had been able to set themselves up in business initially by using their tools. By making the tools soon after arriving at the polytechnic, and then ensuring their use until the trainee left, it was possible to go some way towards overcoming the acceptance barrier.

The need to introduce the skills of toolmaking in youth polytechnics has been recognized by other NGOs working in Kenya. ActionAid, CARE and the International Labour Organisation (ILO), working with the government, set up a pilot scheme in 1983 to see whether it would be possible to pass on the skills of toolmaking effectively to youth polytechnic trainees. A one-month pilot training course was set up to answer this question. The conclusions were positive and, on the basis of this, a long-term programme to train youth polytechnic instructors to teach toolmaking was initiated. This programme ran from 1984 to 1989, and was initially supported by a number of NGOs in collaboration with the government.

ActionAid was central to the programme and alone saw the programme through to the end. By 1989, ActionAid had run courses for selected instructors in all but one district of Kenya. In total, more than 100 instructors had attended courses. At the four-week courses, instructors identified by government officials were taught how to make a basic set of six tools (jack planes, mallets, sash clamps, mortise gauges, benches and vices), designs for which were developed by ActionAid.

Follow-up work showed that very few instructors were putting their skills into practice in their respective polytechnics. ActionAid identified lack of materials and managerial opposition to the idea as two of the main hindrances to the practical application of these skills. Therefore, in all subsequent courses instructors were given materials to take back to their polytechnics. Managers were also invited to attend the last day of

The staff of Gimamoi Youth Polytechnic with some of the tools that have been made there.

the course so that they might be persuaded to accept the idea as useful to their institutions. Problems relating to a low level of skills, motivation and morale were also identified, all of which seemed to be partly to do with the fact that the instructors were not self-selected and therefore did not necessarily have any commitment to the idea of toolmaking. ActionAid, however, had little control over these factors.

At the end of the programme it was felt that the main objectives of introducing toolmaking on a wide scale into youth polytechnics had not been achieved. Toolmaking only really got going in fewer than 10 polytechnics. One of these is Gimamoi Youth Polytechnic in Kakamega District. The reasons for success at Gimamoi are fairly clear as the instructor, who was trained in toolmaking, Mr Peter Burudi, is an extremely well-motivated and skilled carpentry instructor. He is now the manager of the polytechnic which, despite being small, is very well run. In this case all the barriers highlighted by ActionAid had been overcome bar the question of materials. However, Mr Burudi was hoping to set up a small tool production unit to generate funds for materials.

At other institutions, lack of money for materials was put forward as the main reason tools were not being made. At one polytechnic, at Bushiangala in western Kenya, the instructor, frustrated by the lack of materials, had made tools from soft woods but they were not proving durable. There is a lack of funds in the youth polytechnics for materials, and even where ActionAid had given some materials, it did not solve the longer-term problem.

Instructors lack the ability to convince managers of the worth of toolmaking and the one day that managers attended the courses was not sufficient to persuade them. Where managers were convinced and toolmaking was started, poor teaching skills meant there was a loss of skills between instructor and student and the tools produced were of poor quality. The high turnover of staff in youth polytechnics meant that often, just as toolmaking was being accepted, either the manager or the instructor would be transferred and the process would have to start again. The instructors' lack of motivation meant that even where materials were donated, in many cases they were not used. Consequently, the objective of setting up youth polytechnics as centres to supply institutions like primary schools with tools was not met.

Government training centres, Malawi

The very first IT training course in toolmaking was held by Aaron Moore in Malawi at the end of 1986. Four of the ten participants were carpentry instructors from the Salima Youth Rural Trade School, a vocational training centre run by the Malawi Young Pioneers. The others, also instructors, were from the Development of Malawi Traders Trust (DEMATT), the Small Enterprise Development Organisation of Malawi (SEDOM), the Malawi Entrepreneurs Development Institute (MEDI), Soche Technical College, and Glen Forest Training Centre, Zimbabwe. One of the participants, Mr Zinyongo, was a practising carpenter from Tsangano, who is now training Mozambican refugees in carpentry and toolmaking.

Following the first course, Salima and MEDI, the only specialist training institutions at the course who cater for the self-employed, started toolmaking in their own training. Salima Rural Trade School organized a course for some of their former graduates in September 1987, and quickly introduced toolmaking to their syllabus. Graduates from Salima are expected to establish themselves in business in rural areas all over

Malawi, and the tools made by trainees during the two-year course would be taken with them when they leave. At the end of 1987 a set of tools for graduates, which had until then been purchased by the school, cost over 1000 kwacha. Wooden tools made by trainees therefore represented a considerable financial saving for the school.

Blantyre Polytechnic, where the majority of the Malawi's woodwork instructors are trained, took a keen interest in the idea of wooden tools. A course for instructors from the technical training colleges was held there in October 1988. However, it was felt by the instructors that toolmaking could only be successfully introduced to the colleges if the curriculum was formally changed. A seminar to discuss this was held in September 1989, followed by the formation of a working group to develop specific proposals. Assistance was provided by IT and the British High Commission. No changes have yet been brought about as an outcome from this working group.

NGO training centres, Zimbabwe

Glen Forest Training Centre

Both VSO and IT have for several years been working in partnership with Glen Forest Training Centre, an NGO vocational and development training centre situated at Domboshawa, near Harare. Since it was first established in 1980, Glen Forest has been providing carpentry training with the objective of assisting members of rural communities and co-operatives to become self-employed and self-sufficient.

The carpentry training held at Glen Forest now falls into three categories; Basic, Upgrading and Toolmaking. Two or three courses a year are run at Domboshawa and a further two or three courses at their satellite centre, Mwenezi District Training Centre at Neshuru, Masvingo Province. A carpentry instructor from Glen Forest was trained in toolmaking by IT in 1986/7, and since then the centre has, with IT support, held two courses for practising carpenters and three courses for instructors from other institutions.

An evaluation of Glen Forest's carpentry training conducted in early 1990, concluded that the centre itself was as yet unconvinced of the value or potential of self-made wooden tools. (Scott and Moore, 1990) While the centre's two instructors have now both been trained to make tools, the principle of locally-made wooden tools has not been fully adopted. The centre continues to use imported tools for its training, and to separate toolmaking training from other carpentry training. The fact that tool maintenance at the centre was itself a problem, and Glen Forest has continued to call upon IT's technical expertise when toolmaking courses have been organized, says more perhaps about the overall quality of the training and management at the centre.

Silveira House

Silveira House, established in 1964, is a well-known Zimbabwean NGO training centre with a staff of over 90 and many village workers spread throughout Mashonaland. Their training activities include agriculture, nutrition, youth leadership, craft skills, industrial relations, civics and dressmaking. The Skills Department aims to enable people to have greater control over their lives through gaining employment or self-employment. IT has assisted Silveira House to establish a carpentry training programme which includes a toolmaking component. This has included training two instructors in wooden toolmaking and instructional techniques. Silveira House has also been equipped with eight basic kits of wooden tools and benches, all made there, which would have cost twice as much to purchase. The monitoring and evaluation of trainees after training has also been assisted.

In their first year of carpentry training in 1989, Silveira House conducted a nine-month course for three groups of young men, combining blacksmithing and carpentry. Two months had been allowed for the carpentry, to cover basic skills through to toolmaking. Since the carpentry did not allow sufficient time for the making of products during the training, two separate courses are now run for blacksmithing and carpentry. Trainees now use wooden tools from the start of their course and at the end they have a set of tools, which they have made themselves, to take home. During the course they make furniture with the tools.

While it is too soon to judge the general impact of the carpentry training at Silveira House, follow-up visits to date suggest that the young men trained, who were all expected to work as groups, have experienced difficulties in setting up in business and in working together as a group. One of the groups, for example, from Rukodzi Co-operative, received further training in furniture making, but only one member of the group, Charles Nyande, is still there.

St Joseph's

St Joseph's is a small mission training centre at Bikita in Masvingo Province. The centre has, with the assistance of a VSO volunteer, introduced toolmaking to its training. The two-year carpentry course at the centre now has over 40 trainees, and the lack of tools was a major constraint. When Steve Morris, the VSO volunteer, arrived the centre had two old British-made planes and three Zimbabwean planes. Although the centre had been donated tools by the British High Commission, these were mainly of Chinese or Czechoslovakian origin and of questionable quality. Steve Morris had been trained by IT in the UK, and on arrival introduced toolmaking to St Joseph's.

Inyanga Commercial College

Inyanga Commercial College is a small training centre at Inyanga in Manica Province, supported by the District Development Committee, which provides training in commercial subjects, dressmaking and carpentry. In 1989 there were 17 trainees on the three-year carpentry course, most of whom were likely to end up as employees. The instructor from Inyanga, Mr Kamuruka, was trained at the first GFTC course and introduced toolmaking to the centre without further assistance. The supply of hardwood for tools is a problem because it has to come from the other end of the country. They had tried making tools with soft wood, but these were not usable.

Driefontein Mission

A Catholic mission situated 100 km north of Masvingo, Driefontein Mission concentrates on teaching technical and craft skills to school-leavers. The training lasts for two to three years, and most graduates have in the past found jobs in the formal sector. Due to the increasing difficulty of this option, the Mission is now assisting graduates to set up workshops in rural areas. The carpentry instructors at Driefontein were trained in toolmaking by IT, but there has been little contact with the mission since.

Government training centres, Zimbabwe

Ministry of Education
In 1988 five sets of tools were supplied by IT to the Ministry of Education in Zimbabwe, to test in selected secondary schools. Feedback was a long time in coming, but in early 1992 eight woodworker instructors from different provinces were trained in toolmaking by GFTC, with support from IT. This was with the intention that each of these instructors would return to their province and, in subsequent school vacations, train other teachers. The Ministry has also had teachers trained in toolmaking with a view to equipping schools, but it remains to be seen whether the instructors will be prepared to make tools for their schools, and whether incentives and raw materials for this will be provided.

This move has been complemented by the introduction of toolmaking to the course at Belvedere Teacher Training College in Harare, where the great majority of the country's school instructors are trained.

Ministry of Political Affairs
In November 1990, IT (in the person of Aaron Moore) undertook a survey of seven of the youth training centres which offer carpentry instruction, with a view to assessing the problems faced by the centres and by their graduates arising from the high cost and unavailability of hand-tools. These training centres have three-year courses in carpentry and a total of 150 graduates a year. Until now YTCs have provided training to enable graduates to enter the formal sector as employees, and the training includes work experience in formal sector companies.

The survey found that some students had problems with access to tools during their training – there were not enough to go round; that graduates were unable to buy tools, thus limiting their chances for employment and self-employment; and that graduates were not equipped with business skills to enable them to become self-employed. At a couple of YTCs, instructors (expatriates) had attempted to introduce toolmaking but found little support. It was recommended that YTC instructors should be trained in toolmaking and that the curriculum should be altered to make it more relevant to employment or self-employment in the informal sector.

As a result of the survey by IT, and of increased recognition by the parent Ministry that the informal sector is often the only option open to YTC graduates, eight of the 16 YTC instructors have been trained by IT in toolmaking. The Ministry has thus accepted the idea of wooden tools, with a commitment to supply tools to YTCs. Before toolmaking can be introduced, financial provision for the purchase of the necessary materials had to be made and this was expected in the financial year 1992/3. The Ministry also has plans to set up its own tool production unit at one of the youth training centres.

Acceptability

Where the initial problems of funding, time and management acceptance have been overcome, the work of introducing the concept of wooden tools and toolmaking can begin with the students. Getting students to accept the tools has, however, not been easy. The difficulty faced by volunteers in getting tools accepted by their trainees as alternatives to imported tools was clearly shown by the attitudes of a group of six carpentry students at Kianjai Youth Polytechnic in Meru District. These students have all made carpentry tools under the guidance of Andy Ball, a VSO instructor. The question of acceptability in their minds falls into two parts: their own perceptions of the tools and the perceptions of other people (i.e. future customers or potential employees) of them as carpenters using the tools.

None of the students at Kianjai had seen wooden tools before they started making them. Their only images of tools were from text books, the tools in the polytechnic and the tools they had seen other carpenters using. In all instances these were images of imported tools. All the students, when asked, could reel off the advantages of the tools they had made, i.e. 'they are easy to make and repair, they are a little harder to use.'

All six of the boys wished to become self-employed carpenters. None owned all the tools necessary, and they felt that a sum of £400 upwards would be necessary to set themselves up in business, a sum completely out of their reach as none of them had any savings or qualified for a bank loan. They all showed a distinct reluctance even to consider the possibility of setting up in business using wooden tools, which would cost them a fraction of the price. Of those they had made, the only tools the boys were happy to consider using were the ones that looked very similar to bought ones, such as the marking gauge and the mallet, and perhaps the rebate plane as it is not very common. A jack plane was the first imported tool they would buy, and they were even willing to buy the cheaper imported versions although they all knew they were often of poor quality.

It was clear that, although they may have to some extent accepted the tools, a bigger barrier to the boys at Kianjai using the tools outside the workshop was related to the negative attitude they believed that other people would have. When asked if someone would employ them if they only had wooden tools, they all said no. 'They don't understand these tools and don't know if they work,' was one of their comments. All felt that a better alternative would be to find employment in a workshop, using the owner's tools whilst they saved up to buy their own.

Toolmaking: an educational justification

While some of the objectives of introducing toolmaking in training centres may be hard to achieve, many instructors have felt that the process has valuable spin-offs in terms of the education of a young carpenter. Several former VSO instructors felt that their trainees grasped the importance of quality and precision in their work through their experience of making tools. Andy Ball commented,

> 'Often a student's first attempts are poor and they have to start again. However, 50 per cent of the trainees have improved the quality of their work. For example, they really had to sweat making something as simple as a square. If it is not square it's no good. The experience of toolmaking has certainly improved skills.'

Another former carpentry instructor, Ian Marshall, reflected on his experience at Bungasi Youth Polytechnic,

> 'It [toolmaking] created more pride, a little more patience in the students' work and a little more precision. A chair with three legs can basically still be used but a plane or a try square which isn't absolutely correct is no use at all; so making tools emphasized some level of precision in their work.'

All instructors said that the quality of the tools turned out in the end was very high.

As well as quality of work, another skill required to produce more sophisticated end-products is the ability to interpret complex diagrams and designs. Where instructors had used visual aids and offered only the minimum guidance they felt their students' skills in this respect had improved.

Making wooden tools also gives students valuable experience of working with hardwoods, a medium not often used in educational institutions because of its expense. Also, where the instructors have had the facilities, they have introduced their students to some metalworking though this is not common.

Conclusions

In the institutions where toolmaking has been successfully carried out for three years during the posting of a VSO volunteer (such as Kaaga School for the Deaf, Bungasi Youth Polytechnic and Mulango Youth Polytechnic) there is not a single example where the idea, in practice, has outlived the departure of the volunteer. At a number of institutions where toolmaking had been introduced but the volunteer had since left, managers and instructors brought out dust-covered examples of wooden tools from forgotten corners of store rooms and said what a good job the volunteer had done and that if they had the money for materials they would continue. In each case they were buying imported tools to supplement an insufficient stock of tools. Clearly the managers had, for one reason or another, not been fully convinced of the value of toolmaking to their institution, and they were still looking for outside solutions to their problems.

The reasons for this lie in the difficulties of introducing the idea in the first place and then getting the students to accept the tools. Within the fragile financial, managerial and personnel structures of many youth polytechnics and other training institutions, volunteers or external agencies who can bring benefits such as cash donations, and who are often paid from outside the institution, have more influence than their Kenyan colleagues. Local instructors may well have acquired the skills of toolmaking from the volunteer but lack any authority within the decision-making process at the polytechnic and thus are unable to obtain the necessary time and materials to continue where the volunteer left off unless the manager is fully convinced of the worth.

Efforts to bring about changes in the approved curriculum of the educational or training system, rather than at individual institutions, have been moderately successful and many have more lasting effects. The process for achieving this is, however, a long one.

Lessons learned

A number of key conditions have to be met, or positively influenced, before a toolmaking project can be successfully introduced into an educational institution. Unless all these conditions can be met it is unlikely that a toolmaking project will be sustainable.

Materials and funding Materials such as good hardwoods must be available, i.e. the materials must be available in the location and the institution must have a commitment to providing the necessary funds.

Time Toolmaking should be on the syllabus. Where it is not, there must be a commitment from the management and the staff of the institution to integrate toolmaking into the training programme as a part of the curriculum and not as an extra activity. In Malawi the managers and instructors wanted a change in the national curriculum before they introduced toolmaking.

Choice of tools The tools should be simple and relevant to the needs of the local community. The simpler the tool the more likely it is to be copied or made again.

Tool use It is essential that students are taught using handtools right from the beginning of their training. This is the most effective way to overcome the prejudice against such tools.

Skills Toolmaking should be taught in relation to new skills and not as an end in itself: e.g. students should make rebate planes before they begin a project such as a window casement and use the plane to cut the rebates in the frame.

Training instructors An instructor who undergoes training should be self-selected, i.e. have a commitment to toolmaking, be well-motivated (pay and conditions will affect this) and have the necessary skills not only to make the tools well but also to teach those skills to others.

Teaching for self-employment Teaching young trainees to make tools and then expecting them to go into the world of business, where the competition can be extremely stiff, and succeed armed only with their 'pioneering' tool-kit would seem to be unrealistic unless the trainees are extremely determined individuals. The experience of GFTC and Silveira House with training young people in carpentry and toolmaking, and then expecting them to set up in production as a group or co-operative, has not been very successful. The majority of young, inexperienced people cannot realistically be expected to become self-employed. Maybe the best that can be hoped for with respect to the employment objective is that certain tools may help the trainees with the beginnings of a tool-kit, which will increase their chances of employment or self-employment. It would seem even more unrealistic to expect trainees to set themselves up in business making and selling tools whilst the awareness of such tools among fellow carpenters is almost non-existent.

5
Hand-tool production units

Not every carpenter will want to make his own tools. Not every educational institution will want or be able to make tools themselves. The total demand for tools is unlikely to be completely met by self-made tools, particularly if bought tools are considerably cheaper than those already available. Units specially set up and equipped to make hand-tools are one approach to meeting this demand.

Tool production units can help to meet objectives of improving the availability of tools and decreasing their cost. Such units can be of any size, from single carpenters to mechanized workshops, depending on the size of the market to be supplied. They do, however, have to be treated as commercial enterprises if tools are to be made and distributed without subsidies. Experience to date with tool production units has been in Kenya, Zambia and Zimbabwe. In each case tools could be made at lower cost and equal quality to imports, but the units themselves have not been entirely commercial ventures.

Rapogi Youth Polytechnic

In 1989 a tool production unit opened at Rapogi Youth Polytechnic at a rural location in South Nyanza District, Kenya. Although it had the necessary equipment, skills and workshop facilities, within a year it had closed down.

The Rapogi production unit was set up on the initiative of a VSO volunteer, Steve Payne, and the management committee of the youth polytechnic. They aimed to produce a whole range of tools made from metal, for carpenters, mechanics and masons. The equipment needed to start production, such as bench and angle grinders, electric drills and arc welders was purchased and three former trainees of the polytechnic were trained in production skills and taken on as full-time employees. The markets they were aiming at were the artisans and the educational institutions in the area. It was hoped that the unit would generate vital income to help the polytechnic meet some of its training needs and that local markets would benefit from cheaper tools.

Although the quality of the tools was good, they could not sell any despite charging only 15 per cent on top of material costs. Steve Payne suggested that it was not that people did not need the tools, or that people did not want the tools, they just could not afford them. The institutions they were aiming at, such as the local primary schools which had no tools but needed them to fulfil the requirements of the syllabus, were without tools not because they could not afford imported tools but because they could not afford *any* tools. Their budgets for such things were almost non-existent. As for a local artisan market, on closer analysis it was concluded that Rapogi's rural location meant that there were very few artisans in the immediate area. The polytechnic would have to go further afield to find a potential market, something it was not geared up to do. Hence, without a market, the tool unit closed.

Both Steve Payne and the manager of the polytechnic confessed that they had failed to do anywhere near enough market research. Steve Payne felt that the production unit was a good idea but that they tried to put it into practice in the wrong place. To start such a unit the manager recommended carrying out thorough market research, not just addressing questions about which tools to make but also the price the market might pay, and adjust manufacture accordingly. A slightly lower-quality but cheaper tool may have found a market. If more money had been available in the initial stages it may have been possible to withstand a slow market and they might have had the resources to look for markets further afield.

Bande Tool Makers

Bande Tool Makers, a small-scale enterprise run by Evans Njeri in Meru town, produces wooden carpentry tools. Bande Tool Makers were given help to set up in business by the new NGO, Approtec, in a bid to determine how successful toolmaking in the private sector might be. Evans Njeri employs several young carpenters to assist him, as the need arises. He has been subsidized quite heavily by Approtec with the renting of a sizeable workshop, with electricity, and the supply of some of the materials he needs. ActionAid have placed an initial order of 60 tool-kits with Bande, which they will donate to targeted primary schools with the aim of spreading awareness of such tools.

Evans Njeri was trained by ActionAid and subsequently attended a course at Kianjai Youth Polytechnic as part of the YTSP programme (mentioned in chapter 3). Evans has acquired a high degree of skill in toolmaking and is capable of making a whole range of quality wooden tools. The tool-kits he is producing for ActionAid are made up of six tools in a wooden box including a jack plane, mallet, sash clamp, square, sliding bevel and mortise gauge. The box and the tools are being sold to ActionAid for £30, which is about 30 per cent of the cost of the imported tools, and which gives Evans 15 per cent profit. He is using designs supplied by ActionAid with a few alterations to make them more suitable for young children. VSO has also placed an order with Evans for 15 jack planes and mallets and 10 sash clamps. He has only made one or two tools for local artisans, and it would be true to say that as yet Bande Tool Makers have been supported solely by NGOs and that the strength or weakness of the commercial market has not been fully tested.

Evans has had no problem getting appropriate materials, Meru being the heart of the hardwood-producing area of Kenya. Blades have been purchased from WECO. Maintenance of quality has, however, been a problem. The quality of a handmade wooden tool has to be very good if it is to work well and have any hope of acceptance. When producing such large numbers of tools by hand, quality control is a problem, especially when finding artisans trained in toolmaking is very difficult. Thus some of Bande's tools have been of a poor quality. This problem may be exacerbated when the NGO orders dry up and Evans has to leave the workshop for longer periods of time to market the tools in the local area.

Evans recognizes there will be difficulties in finding a local market for his tools, but he feels there must be a market in local educational institutions and amongst local carpenters, though no market research has been done. He is aware of the fact that, should this market exist, he will need to spend a lot of time demonstrating the use of the tools to his customers, time that will have to be charged in the pricing of the tools if he is to be profitable.

In many ways Bande Tool Makers are still untested in the real market and whether or not their business can be viable on such a small scale has yet to be proven. Bande Tools have not established a real market price for the tools and Evans' entrepreneurial skills have not been tested. The chances of a single small-scale producer surviving on their own, having to do the necessary marketing, awareness-raising, training, instructing customers in tool use and maintenance, and assessment of local needs all on a 15 per cent profit margin are slim. Maybe in fifteen years time, when the tools that are being given to primary schools have given rise to a new generation of carpenters who have accepted the worth of such tools, Bande Tools will be viable.

WECO tool production unit

In June 1988, David Boothman a VSO volunteer, was placed at the Western College of Arts and Technology (WECO) in Kakamega. This is a large, well-equipped, national, technical training centre. The project at WECO supported by VSO was to establish a tool production unit which could provide capital for the college. Another production unit, run by Danish volunteers making water pumps, had previously been set up in the college and so the machinery necessary for tool production was already there. The first task was to design tools, from both wood and metal, that could be made in the workshop, under quality control, by a workforce of around 10 former WECO trainees. The initial objectives of the unit were to produce tools of a high quality for use in local educational institutions but the unit has also produced tools for local carpenters, masons and metalworkers. The aim was to produce tools of a better quality than the cheaper imports but at a price that could challenge the more expensive ones.

Since 1988, WECO has produced 65 different designs of tools ranging from wooden smoothing planes through metal vices and benches to masonry chisels, although not all have come into production. David Boothman handed over the unit in August 1990 to another volunteer whose skills were more directed towards production management. It was hoped that his skills would help to put the designs into production in large numbers. Initially more time was spent on the design process than the production and marketing of the tools. As the range of tools

Members of the Kakamega jua kali *try out the metal WECO jack plane.*

to be made were all of new designs it was crucial to get them right. It was very important to create tools that would be of use to the local market. For example, the idea of designing sash clamps and vices had come from people bringing him cheap imported examples that had broken. David Boothman felt it would be possible to make a far more durable version for the same price or less.

By August 1990 the production unit was sustaining itself, though not making great profits, but it was being mainly supported by an artificial market created by NGOs and educational institutions. The order-books were full, however, and the problem facing the unit was not lack of buyers but an inability to get enough money together to acquire a large stock of raw materials. A second problem was that they were not being allocated enough time on the necessary machinery, as the tool unit had to share the machines with the college trainees.

The need for marketing by WECO became clear from a visit to the Kakamega *jua kali* sites, less than two kilometres from the college. Here none of the carpenters had any idea that carpentry tools were being produced in the locality. Apart from displaying the tools at national and some local trade fairs, little, if any, marketing had been done. Although WECO tool production has been propped up by an artificial market there may well be a local market for the tools, in particular for the metal ones as they are very similar to the imported ones. A lot of interest has been shown in the WECO tools by the *jua kali* artisans who have come into contact with them. It was recognized that there is a need to concentrate production on a few of the more popular tools like the plough planes, sash clamps, jack planes, etc. and build up a stock that can then be offered to the *jua kali* and educational institutions.

To do this, problems of machine time and cash flow need to be sorted out. It was noted that money management within the college was not one of its strong points and that one of the difficulties of placing the unit, or any production unit, within an educational institution was keeping track of the accounting, to make sure that production unit money did not go elsewhere within the college. The necessary managerial, technical, financial and production skills were not present in the college to sustain such an enterprise. However, the objective of supporting the establishment of the WECO unit was to prove that toolmaking of a quality and price to compete with the imported tools is a viable economic venture, and not to prove that such a unit could be sustainable within an educational institution. WECO tools are there to create a market for private entrepreneurs to follow.

Itawoga tools

The preliminary investigation of the potential for wooden carpentry hand-tools in Zimbabwe, carried out for IT in 1987, found that the market which does exist would be unlikely to accept the quality of tools made in small workshops, but that high-quality tools, manufactured in a production unit, could well find a market in Zimbabwe's educational system.

Ten sample sets of wooden tools were then produced by IT at the workshop of the Danhiko Project on the outskirts of Harare. These included a jack plane, smoothing plane, rebate plane, plough plane, mortise gauge, marking gauge, mallet and try square, and were tested in schools run by the Ministry of Education and the Zimbabwe Foundation for Education with Production (ZIMFEP).

Then, in 1988, ZIMFEP in collaboration with the University of Zimbabwe's Development Technology Centre, carried out a detailed feasibility study for a hand-tool production unit. Their market survey found a conservatively estimated demand for 277 kits in the first year, rising to 479 kits (or 3832 tools) by the fifth year. (Gross and Lum, 1988) The bulk of this market was in the educational sector, with 188 kits at training centres and 97 kits in schools. The feasibility study found that the supply of carpenters' hand-tools was not adequate to meet demand and the capacity for a production unit was placed at 450 kits per year.

Following the feasibility study, ZIMFEP and Danhiko decided to establish two units, one at Danhiko's existing workshop and the second at a ZIMFEP-supported co-operative, Grow More Trees, at Chegutu. In June 1990, IT provided training for two members of Grow More Trees and three Danhiko employees in the design and construction of jigs and fixtures for tool production. Though this training had been requested, it was not felt by IT to be entirely relevant for the task in hand, i.e. establishing a production unit. IT also assisted with the setting up of the production system, and initial monitoring of quality. Production was successfully established at Danhiko, which had most of the mechanized equipment needed as well as experience in carpentry production.

Production levels have so far been considerably lower than had been anticipated by the feasibility study, the result of a conscious decision of Danhiko to proceed cautiously. Grow More Trees has to date concentrated on meeting orders for furniture. The tools are now to be jointly marketed under the brand name Itawoga Tools and, as well as selling via direct orders, some sales have been made through retail outlets. The workshop at Danhiko has faced difficulties with the supply of seasoned hardwoods, which have to come from the south west of the country. In addition to erratic supplies of hardwood, production is interrupted by training activities being held in the same workshop.

More recently, the government has taken an interest in establishing a mechanized tool production unit at one of the youth training centres. Such a unit, with an initial capacity for 200 tool kits a year, would be for the supply of wooden tools to the youth training centres. This would be instead of buying them from Danhiko and Grow More Trees.

COMET, Zambia

A wooden hand-tool production unit has been established at Kabwe by COMET (Copper Mine Enterprise Trust), an organization founded by Zambia Consolidated Copper Mines Ltd. The unit was set up with a view to providing direct employment and self-employment opportunities for carpenters. Establishment of the production unit was preceded by a feasibility study which optimistically estimated that there existed a market in Zambia for almost 1600 tool-kits a year. This demand, based on a finding that there are some 18 000 practising carpenters in the country and 400 training centre graduates a year, would be constrained by purchasing ability. The feasibility study found that plough and rebate planes were unavailable in Zambia and estimated that jack planes made locally would be a third of the price of imported planes.

The production unit was established in 1990, with the capacity for 500 tool-kits a year. Each kit of eight tools would contain a jack plane, rebate plane, bench plane, plough plane, try square, mallet, mortise gauge and bevel square. The unit, using an existing building, required importation of machinery (sanding machine, dimensional saw bench, overhead router, thicknesser, surface planer and dust extraction equipment), at a total cost of 342 335 kwacha (£26 000). The total capital cost of the unit amounted to over 555 000 kwacha (£42 000), and would employ six workers, though the major benefit was not intended to be direct employment but availability of lower-cost, local tools for other carpenters and training centres.

IT provided advice to COMET on setting up, procurement of equipment, training of the workforce and manufacture of jigs for production. Though the unit experienced management and staffing difficulties, by early 1991 it was felt to be ready for full production. The unit also experienced problems with equipment maintenance and the need to import parts, though it is difficult to separate these from the management questions.

The Trades Training Institutes of Zambia, which offered a potential market for 50 kits a year according to the COMET feasibility study, have now embarked on training their own staff to make tools with a view to introducing them to their courses. This might affect the total market, but may not really be critical to the continued performance of the unit at Kabwe.

Conclusions

These examples of tool production units clearly demonstrate the need to approach such projects in the same way as any other business venture. Rapogi's failure may not have had anything to do with the tools themselves; necessary basic groundwork had clearly not been done. Rapogi produced the wrong product in the wrong place. Bande Tool Makers, as yet, have not had to confront a 'real' business environment. They have proved that tools can be produced at a fairly competitive price, but have not asked the vital questions that any business should: i.e. Who are our potential customers? How do we effectively reach those customers? Is our product the right one for the market? How could we best market our products? etc. It is very unclear whether Bande Tool Makers will find satisfactory answers to some of these questions. WECO, on the other hand, has got answers to some of these questions. For example, designs have at least been produced in response to local needs and market demands. The WECO product may well suit the market but WECO still has to see if its potential customers can be persuaded to change allegiance from an imported metal plane to a WECO metal plane, which is a small step but may prove a difficult one.

Like Bande Tool Makers, WECO has still to prove itself in the local market. Although WECO has a small percentage of its sales within the local market, should the artificial market disappear, the survival of the production unit is questionable. Given the problems of acceptability that locally-made handtools suffer, WECO's way into the market is going to be very difficult. With some strong marketing of the metal tools their future is perhaps better than that of Bande Tools as the WECO tools are more 'recognizable'. Many artisans shown the WECO tools doubted whether they were made in Kenya. It may even be possible, if production is stepped up, to offer the tools at an even more competitive price. Once a few of the tools have been accepted and they gain some credibility, the advantages of price and durability may well help them to overcome the prejudice of a wider market.

In Zimbabwe and Zambia the approach has been more businesslike, with feasibility and market studies, but this has not guaranteed the success of the production units. The viability of a tool production unit, making wooden tools for a market whose awareness of such tools is almost non-existent, is extremely low, as suggested by previous chapters. For example, after purchasing a blade and the materials, then paying labour and overheads, Bande's wooden jack plane retails at £8. An imported Stanley jack plane can be bought for around £25. Whilst this is a considerable difference, the saving is not enough given the limitations of the wooden jack plane and the prejudice against it. The attraction of a wooden jack plane to a carpenter is surely that he can produce it himself for around £3 or £4 at his own workshop or institution. The WECO metal jack plane, which costs £9 and has advantages of durability over its cheaper competitors but is still under half the price of the more expensive imports, may well prove a more successful product.

The prices of the WECO tools may not be realistic, however, if they were to be produced in the private sector. Although the workforce at WECO is paid a salary, it is a little lower than an equivalent formal sector wage. Employment at WECO is seen as a stepping-stone to the formal sector, a kind of apprenticeship. The machines and workshop at WECO have been provided free of charge; therefore the overheads are very low. If a completely independent entrepreneur in the private sector took to producing similar tools, the price would probably be less competitive.

It has been noted that as yet no independent entrepreneur has taken up the challenge of setting up a toolmaking business. Evans Njeri was groomed and aided by Approtec and ActionAid and perhaps lacks the aggressive entrepreneurial drive to meet the challenge. Several entrepreneurs spoken to considered the hand-tool market to be completely dominated by the large imported tool businesses and that it would be far too risky to try and challenge them.

Lessons learned

Programme planners need first to be clear about their objectives when considering the establishment of a hand-tool production unit. For example, is the role of the unit to demonstrate that local tools can be viably produced (if so, experience so far in the semi-protected NGO environment has not proved this) or is it to produce tools that are otherwise unavailable? Different objectives inevitably determine the ways in which a project will be implemented and the ways in which success and failure can be judged.

Market research Before starting a tool production unit, thorough market research must be carried out to address the questions of 'what tools, what quality, what price?' etc. Omitting to do this can cause immediate failure, as in the case of Rapogi. Perceptions of quality and the desirable characteristics of hand-tools need to be understood as part of this market research in order to assess the potential acceptability of the tools.

The range, quality and design of the tools to be made should be determined in the light of market research. In those cases where the specific needs or even the initiatives of local carpenters in producing tools can be identified, production units may follow the lead of the informal sector by improving the standard of locally available tools for which there is a proven demand. It is important initially to concentrate on a small range of tools in order to keep up production levels, aid marketing and keep prices low.

Location The location of the unit should be given careful consideration in the light of the objectives and market research. Urban centres are likely to have a larger and more diverse market. The availability of raw materials (e.g. hardwoods) is another factor determining the suitability of the location.

Capital There must be sufficient initial funds available to start and sustain production. It is important to have enough working capital to create a stock of tools. One of WECO's problems is that it can only cater for orders. This limits its market as local artisans are likely to want only one tool and to want it straight away.

Financial control Where a production unit is to be set up in an educational institution, strict independent control should be kept of the finances.

Quality control Given the problems of acceptability faced by locally-made tools, any loss of quality will have serious repercussions on the confidence of the potential market. Quality relies on having a stable, skilled and well-trained workforce. Jigs and fixtures have been used at Danhiko and COMET to ensure consistent quality, but these represent a substantial investment and are only warranted for regular and volume production.

Marketing Marketing of the tools is very important. Tools should have an attractive finish that is identifiable with the unit in order to enhance their consumer appeal. A brand name, such as Itawoga Tools, could help. The tools should be displayed and well advertised within the location.

6
Conclusions and future work

Skills transfer

Experience has shown that low-cost, good quality, durable tools can be produced simply, from locally obtainable materials. It shows that transfer of the skills necessary to teach others to make their own tools is possible. The experience of VSO and IT working with trainee carpenters and more experienced carpenters in the informal sector, shows that with the right approach to teaching methodology, the skills can be passed on effectively and quickly.

Appropriateness

Carpenters of the Kisumu *jua kali*, and elsewhere, clearly demonstrate that tools will be, and can be, made and used by carpenters themselves under conditions where the idea is accepted and the need for tools is sufficiently strong. The ultimate test of the appropriateness of locally-made tools is whether or not the tools are used. In the Kenyan experience the examples where the tools have been successfully utilized, in both educational institutions and private enterprise, are few but they do demonstrate that where prejudices attached to these tools have been successfully overcome, the tools can be made and used effectively. The benefits that these tools can offer to carpenters in developing countries make them an appropriate technology in such contexts.

Acceptability

Acceptability is the biggest single factor preventing the widescale use of locally-made hand-tools and will be a major problem for any programme of toolmaking where imported tools are available. Although examples of carpenters who have accepted local tools in Kenya are limited, in the face of the modern marketing accompanying imported tools, such a small step should not be underestimated. As Aaron Moore has noted, 'Stanley planes, like Coca Cola, have been so well marketed in the developing world that no substitute will do.' (Moore, 1991)

Examples of methods to aid acceptance in the different areas have been documented. In the end, whether one is asking a carpentry trainee to make and use tools or is trying to sell a tool to a carpenter in the informal sector, neither is going to accept the tools happily until their benefits have been clearly demonstrated. Therefore, any programme must address the issue of 'selling' the idea, and include strategies to overcome prejudice. The problem of acceptance may become smaller as the price of imported tools rises, but it will never disappear. It must be remembered that introducing locally-made tools is not just a question of challenging perceptions of what a tool is, but also confronts the stronger perception that imported, factory-made products are best. In Kenya it is certainly proving to be a hard task, as it surely will be in many countries, but inroads can be made and it may be possible to achieve a critical mass of local hand-tool presence which will lead to eventual success. Once a carpentry trainee sees a local tool being used in the 'real' world of the carpenter's workshop he is more likely to accept the tool himself.

Sustainability

As yet the experiences drawn from Kenya demonstrate that it is extremely difficult to build sustainability into a toolmaking project. Sustainability is closely linked to acceptability, with implications for the market, and thus the success of any project needs to be seen in this context. Sustainability of making tools in a production unit, or as part of an educational institution's curriculum, will only be achieved where there is an acceptance of such tools, so any strategies to aid acceptability should enhance sustainability.

Combined programmes

Where toolmaking can be introduced by various approaches and at different levels within one region, sustainability through acceptability could be improved. By introducing tool-making into educational institutions and the informal sector and getting the idea accepted by influential government officials, all in the same region, it may be possible to influence perceptions from several different angles. The process would be less dependent on an individual. The tools become familiar objects that have value, that can be found in educational institutions, technical training centres, carpenters' workshops and can be purchased from the local tool production unit. In this way individuals using local tools would no longer be pioneers of a strange idea, but part of a wider process.

Which tools?

Whether a toolmaking programme is to be aimed at educational institutions, is introducing toolmaking to local artisans or is setting up a tool production unit, the tools to be made should be directly related to the needs of the consumer. The experience in Kenya and Zimbabwe has shown that more use is made of locally-made specialist planes than locally made jack planes, and that tools which resemble the imported versions (for example, gauges) are more readily accepted. Each country and even district within that country may well have different tool needs. In all cases close examination of tool need in each area should happen before specific tools are introduced.

Self-employment

One of the primary aims of introducing toolmaking in Kenya, Zimbabwe and Malawi was to enhance the ability of young carpenters leaving technical training institutions to enter self-employment. This aim has not been achieved to any great extent. Although lack of tools is certainly a factor influencing the ability to become self-employed, there are also many other constraints that need to be overcome before success in this regard can be achieved. Lack of experience, basic business skills, capital and a suitable workplace are all problems that have to be overcome before a young carpenter can hope to set up a successful business. A young man will also lack status in the eyes of the community, a fact that will further hamper his chances of successful self-employment. The majority of self-employed carpenters have spent several years working for other carpenters, in either the formal or informal sector, before setting up on their own. In doing this they have gained experience, some capital with which to buy tools, and have picked up some rudimentary business skills. Therefore, any toolmaking programme which is seeking to help carpenters into self-employment should incorporate elements to help tackle these other areas. It may also be advisable when working in the informal sector to target those age groups which have already gained some experience.

Further work by IT and VSO

From Aaron Moore at Kaaga School for the Deaf in 1983, to the more extensive efforts of IT and VSO during the late 1980s, the promotion of toolmaking has been pursued by agencies and individuals convinced of its relevance to carpenters and to the unemployed youth of Africa. A great deal has been learnt about how to introduce toolmaking, and about the pitfalls that might be expected. Evidence of the value of making low-cost tools has been gained. Based on this experience, both VSO and IT are planning to continue their support of toolmaking programmes, adapting their input to respond more closely to the needs of the target groups.

A review of IT's programme in Zimbabwe, in 1991, concluded that in a relatively short space of time there had been some success in disseminating the concept of locally-made wooden hand-tools. The programme is to continue for at least another three years, with a focus on establishing toolmaking in the state secondary education system and in the youth training centres. This will entail the training of instructors and the development and amendment of training curricula. At the same time support to the work of NGO training centres will be provided. The work in Zimbabwe has generated interest in neighbouring Zambia and Mozambique, and programmes of work with organizations in these countries may be developed.

VSO Kenya's toolmaking input data has been largely centred on the work of individual volunteers posted to individual technical training institutions. This has proved unsustainable and VSO with the British Overseas Development Administration and the Ministry of Technical Training and Applied Technology have initiated a project that will attempt to synthesize and learn from the experience of the previous eight years and adopt a new approach to the problem of lack of tools in Kenya.

The VSO-ODA project is the construction and establishment

of a district training centre at Kianjai Youth Polytechnic, situated about 20km from Meru. The centre will run residential, five-day courses for youth polytechnic trainees, their instructors and *jua kali* artisans. In an attempt to identify and respond to local problems and needs from the start, the courses in the first year will cater for *jua kali* carpenters. These practising artisans will have the option of instruction in marketing and business skills, areas of need that were neglected in previous initiatives. The participants will also learn improved techniques in making, fixing and finishing both hand-tools and finished products as well as product diversification. Regular evaluation of the project will take place to adapt the nature of the course to the requirements of the carpenters. If successful, the project will be replicated in other districts.

In these ways VSO and IT aim to improve the effectiveness of their support to carpenters, responding more closely to local needs and changing conditions in local markets. Central to this is an understanding of the dynamics of small-scale enterprises and of the informal sector.

References

Abuodha, C., 'The needs of the *jua kali* sector' in *Report on the Jua Kali sector workshop*, VSO Kenya, Nairobi, (1991).

Cromwell, G., and Moore, A., *Rapid assessment of rural carpentry enterprises in Zimbabwe*, Report for ITDG, Rugby, (1988).

FAO, *Small-scale forest-based processing enterprises*, FAO Forestry Paper 79, FAO, Rome, (1987).

Gross, G.H., and Lum, R., *Feasibility study for the production of low cost wooden woodworktools in Zimbabwe*, ZIMFEP/DTC, Harare, (1988).

Haggblade, S., Hazell, P., and Brown, J., 'Farm–nonfarm linkages in rural sub-Saharan Africa', *World Development*, 17 (8), 1173–1201, (1989).

Livingstone, I. 'A reassessment of Kenya's rural and urban informal sector', *World Development*, 19 (6), (1991), 651–670.

McPherson, M.A., *Micro and small-scale enterprises in Zimbabwe: Results of a country-wide survey*, GEMINI Technical Report 25, DAI, Washington, (1991).

Moore, A., *How to make twelve woodworking tools*, IT Publications, London, (1986).

Moore, A., *How to make planes, clamps and vices: Seven woodworking tools*, IT Publications, London, (1987).

Moore, A., 'Do it yourself: Carpenters' hand tools', *Appropriate Technology*, 18 (2), (1991).

Moore, A., *How to make carpentry tools*, IT Publications, London, (1992).

Moore, A. and Scott, A., *Report of an evaluation of carpentry training*, mimeo, ITDG/GFTC, Harare, (1990).

Mutiso, J., 'An experience', in *Report on the jua kali sector workshop*, VSO Kenya, Nairobi, (1991).

Nkungi, B., 'Policies and guidelines', in *Report on the jua kali sector workshop*, VSO Kenya, Nairobi, (1991).

READI, *New directions for promoting small and medium scale enterprises in Malawi: Constraints and prospects for growth*, Malawi/USAID, Lilongwe, (1987).

Scott, A., *Review of the carpentry hand-tools project in Malawi*, Report for ITDG, Rugby, (1987).

Woodworker, June 1991. This issue contains details of Steve Morris's activities (see page 31).

Appendices

APPENDIX 1
The tools

This book cannot begin to describe the designs and methods of making carpentry hand-tools. This is best left to others and a number of publications are available which do this already (e.g. Moore 1986, 1987, 1992), in ways which the specialist reader would appreciate. This appendix describes some of the tools which have been discussed in the text, providing more information for those interested in the tools themselves.

Jack planes

There are a variety of designs for wooden jack planes, the most common tool and the one which all carpenters would buy first. The stock can be made from one piece of hardwood, from two pieces glued, screwed or held together with dowels, or from three pieces (one piece with two side pieces). The last seems to be the most popular. The blade is held in place by a wooden wedge which locks tightly in place between the blade and a wooden crossbar. Different handle designs can be added and the plane can have sharp or rounded edges. (See photographs 1 and 2)

Wooden jack planes are some of the most complicated tools to make, along with other types of plane. When well made, however, they are capable of producing work of high quality and they are durable. The main criticism of this tool is that it takes time to set the blade.

Metal jack planes have been made by WECO in Kenya. The stock is made from three-inch cold rolled channel, 4mm thick. Angle iron is used to make the support for the blade which, along with the chip breaker, is held in place with a bolt. Arc welding equipment is needed to assemble the plane, and the aperture is made using a slot drill in a vertical milling machine. Handles are made of wood.

Setting requires the use of a spanner to adjust the bolt

and although it takes a little longer at first, with practice it is easily mastered. The plane is very durable and does not shatter when dropped, unlike many cheap imported planes. (See photograph 3)

Rebate planes

A wooden rebate plane can be made with or without a fence, though the version with a fence seems to be more popular. Photograph 4 shows a rebate plane that uses a chisel as a cutter which is held in place with a wedge.

Rebate planes, particularly the one shown in photograph 5, proved popular with those who made it on a YTSP course and has potential because all imported versions are very expensive and sometimes scarce.

Plough planes

Wooden and metal plough planes have been made successfully and both have proved popular due to the expense and scarcity of imported versions. Wooden plough planes can be made to use a chisel as a cutter, but they do need some metal parts (e.g. sole plate and guide). (See photograph 6)

Metal plough planes can be made from a one-inch square mild steel bar with angle iron for the guide frames and round metal bar for the guide handle. (See photographs 7 and 8) Arc welding and drilling equipment are necessary.

Clamps and vices

Wooden clamps and vices use wooden wedges to apply pressure to the workplace. They are relatively easy to make and are much cheaper than imported versions. It takes a little time to learn how to use them, but once mastered they have proved very effective. Many carpenters spoken to could not afford the number of sash clamps they required, so these could prove very popular.

In Kenya, WECO produces metal G-clamps, sash clamps and bench vices (see photograph 9), all of which look very similar to imported versions. They require arc welding, drilling and other machinery to make. Their cost is comparable to the cheaper imported versions but they have the advantage of durability.

They have proved quite popular with artisans in Kenya. Photograph 10 shows a wooden G-clamp.

Marking, measuring tools and mallets

Making these tools from wood is relatively simple and, as many of the imported equivalents are also wooden, acceptability is not a problem. Many carpenters do not buy marking and measuring tools but improvize their own versions.

Imported equivalents are very cheap, therefore local production cannot easily compete. However, these are good tools to make with trainees who are making tools for the first time because they are relatively simple and introduce toolmaking techniques. (See photographs 11, 12 and 13 for wooden versions of these and photograph 14 for the range of measuring and marking tools made by WECO)

Chisels, screwdrivers and other metal tools and parts

Although metal chisels, screwdrivers and plane blades can be successfully made using scrap metal, the temperature control of the forge and heat treatment techniques are an acquired skill needing precision. It would therefore seem more sensible to teach these techniques to metalworkers, who are already used to the processes involved, than to try to teach them to carpenters. (Photograph 15 shows a range of chisels, screwdrivers and a hammer made at WECO)

1. Jack plane made by Bande Tool Makers. An ActionAid design.

2. Jack plane. Made by YTSP trainees. Design by K. O'Donnell, a VSO carpentry instructor.

3. Adjusting the metal WECO jack plane.

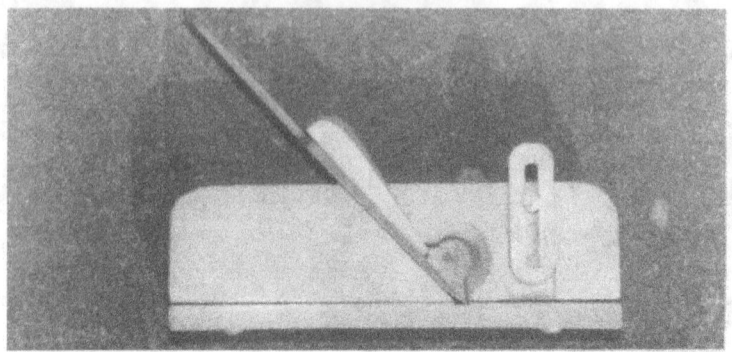

4. Rebate plane made by students at Kianjai. Aaron Moore design.

5. Rebate plane made by students at Gimamoi. ActionAid design.

6. Plough plane. Aaron Moore design.

APPENDIX 1

7. *Plough plane. Made in Kisumu* jua kali.

8. *Plough plane. Made at WECO.*

9. *A collection of clamps and vices made at WECO.*

APPENDIX 1 67

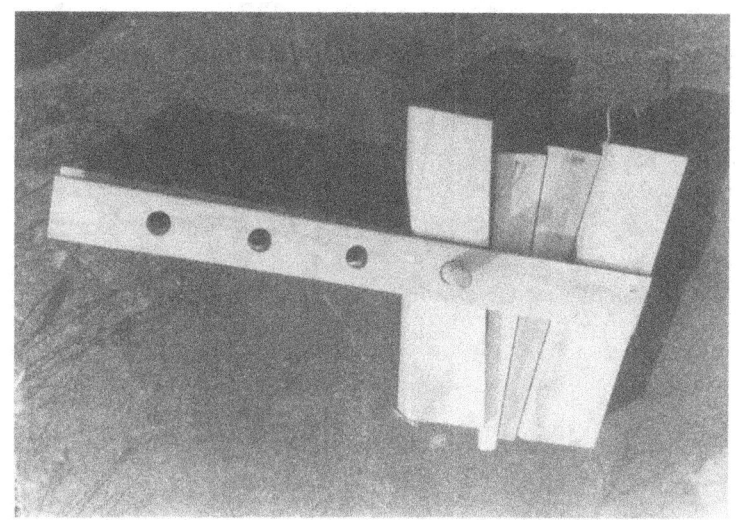

10. Wooden G-clamp made at Kianjai. Aaron Moore design.

11. Mallet. Made by Bande Tool Makers.

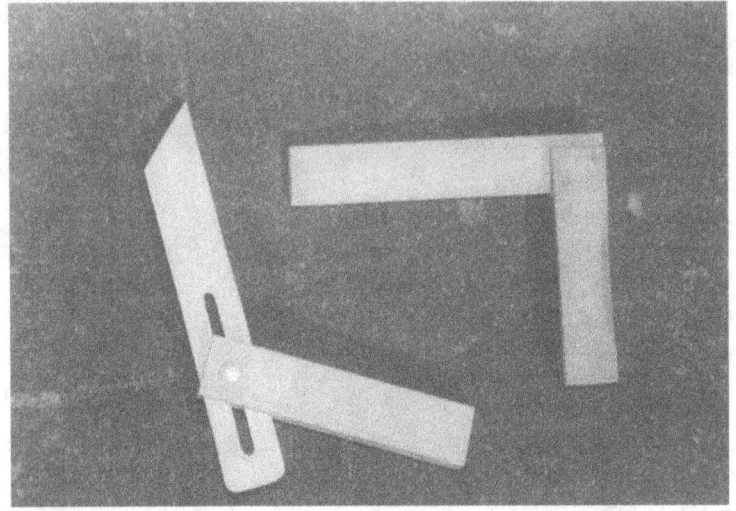

12. Sliding bevel and square. Made at Kianjai.

13. Mortise gauge. Made by Bande Tool Makers.

APPENDIX 1

14. A range of measuring and marking tools made at WECO.

15. A range of chisels, screwdrivers and hammers made at WECO.

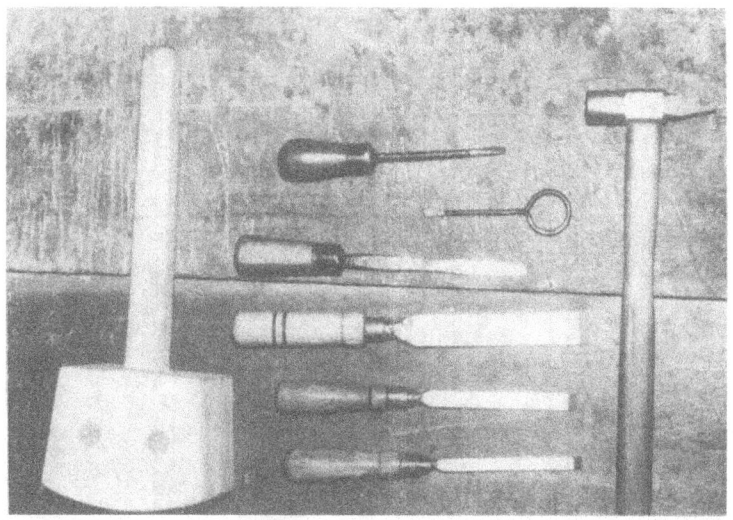

Appendix 2: Price comparisons of selected tools

Tool	Kenya KSh. (1991)			Zimbabwe Z$ (1991)		Malawi Kwacha (1987)	
	Imported	WECO	Bande	Imported	Self-made	Imported	Self-made
Jack Plane	900	450	400	191.91	50.50	140.00	21.00
Rebate Plane	1150	150	–	150.00	34.50	131.75	15.00
Plough Plane	3650	450	–	180.00	52.90	550.00	24.00
Try Square	450	120	50	16.31	11.50	6.95	1.00
Sliding Bevel	720	–	80	47.00	16.75	–	2.00
Marking Gauge	425	80	–	45.00	15.00	13.50	2.50
Mortise Gauge	500	–	85	–	–	27.50	3.50
Sash Clamp	2800	550	275	172.45	57.10	118.50	5.00
Mallet	–	65	90	35.00	22.70	–	3.00

Exchange Rates with the Pound (£): KSh.50.00 (1991); Z$ 5.00 (1991); K.4.50 (1987)

The ECOE Programme

(Evaluating and Communicating our Overseas Experience)

The Need

Over the past thirty years, more than 20000 volunteers have worked abroad with VSO. Currently, there are over 1500 volunteers working in over 40 developing countries in Africa, Asia, the Pacific and the Caribbean for periods of two years or more. However, we have become increasingly aware that much of this valuable experience has been lost through not being recorded in ways which make it accessible and communicable. The ECOE Programme addresses this problem.

The aim

The aim is to record volunteers' experience in reports, videos, seminars, conferences, books, etc. This body of knowledge supplements and supports the work of individual volunteers. It also provides information which is accessible not only to volunteers but also to their employers overseas and to other agencies for whom the information is relevant. Care is taken to present each area of volunteer experience in the context of current thinking about development so that VSO both contributes to development discussions and learns lessons from them for the continuance of its work.

Advisory panel

A panel of opinion leaders in relevant professions and in development thinking advises on the selection and commissioning of ECOE publications.
For further information write to:

The Programme Evaluation Manager
VSO
317 Putney Bridge Road
London SW15 2PN, UK
Tel: 081-780 2266   Fax: 081-780 1326

www.ingramcontent.com/pod-product-compliance
Lightning Source LLC
Chambersburg PA
CBHW071029080526
44587CB00015B/2542